INTRODUCTION

There is a common misconception regarding home construction, that you must have extensive knowledge of homebuilding to do it yourself. The truth of the matter is you really don't need a lot of construction experience; you simply need to be very organized and be a good communicator & coordinator. Attention to detail is also very helpful.

Most contractors will add a percentage fee to each of the sub-trades working on the project as well as a marked up dollar amount on materials. By acting as your own contractor on the project you should save all these fees and more. By being organized and coordinating all the work, you can save time which amounts to further dollar savings by shortening the term of your construction loan

In the following pages I will show you step by step the order of construction trades, how to hire dependable –qualified workers, the do's and don'ts of subcontract work agreements, and how to protect your interests by having the proper insurance coverage throughout the entire project. This publication takes you from the concept stage of your project through the foundations, framing, roofing and thermal insulation.

It is my intent, that through this publication, you will greatly benefit from the valuable lessons and secrets I learned in my 35 years of constructing single family homes, multifamily resort & hotel properties and various commercial projects. Valuable secrets from previous projects that will help make your construction project an enjoyable and profitable experience.

TABLE OF CONTENTS

Finding the Right Property

A few things to consider when looking for the perfect property on which to build your new home. Choosing a building site inside the city or in the country both have unique pros and cons. For example, if you decide to build in the city, more than likely you will choose a building lot in a development which already has utilities located on or near the property which is a good thing. You may want to consider how close you are to work, schools, shopping, transportation and medical facilities depending on the needs of your family. However, there are some things to watch for when choosing a building site inside the city limits;

If you choose a lot in an established development, there may be a Homeowners Association which not only charges a monthly or annual fee, they may also have restrictions on the size of home, how many stories, minimum square feet, type of exterior finishes and colors allowed, and other restrictions. In addition to the HOA, you may have to submit construction plans and specifications to the local building authority for review and permitting, and be required to pay permit fees prior to starting your project. In this case, you will also be required to call the building authority to inspect your construction project at different intervals to insure the project meets all the minimum requirements of the local building codes. Some areas are taxed by both the city and county so you will need to check the specifics before you decide to buy in an area within the city limits.

If you choose a building site in the county or outside the city limits, you may not have the same permitting and inspection requirements. You will need to check with your local building authorities in each area to see just what the requirements are. Also, there may not be a homeowners association to govern the size or type home you can build. While this lack of restrictive guidelines seems great, consider that the man building on the property next to you has the same freedom to build whatever he wants also, which could work against you in the long run. Building in the county can also have a unique set of challenges when it comes to having the utilities you need. Many areas are still undeveloped and don't currently have electricity, water, sewer, gas and cable readily available to your property. In this case you may have to pay extra for one or more of these utilities to be extended to your property for the project. Do your homework and check this out with the local county offices before purchasing the property.

Lastly, check with the local government offices to determine if you are in a flood zone.

Choosing a Plan

With so many house plans available on today's market, how can one decide? You will need to take your time with this one. To start with you will need to list the items that your family sees as a priority when it comes to a home plan. For example, consider;

- How many bedrooms do we need currently, and in the future?
- Do we need a single story or prefer a multistory home?
- Do we spend a lot of time in the kitchen preparing meals? How large of a kitchen and what type of appliances do we want to include? Gas or electric?
- Do we prefer all the bedrooms on the lower level or some upstairs?
- Do we prefer the master bedroom on opposite sides of the home from guest rooms?
- How many bathrooms does our family need?
- What about storage areas? Typically this gets shortchanged on many plans
- Parking areas and drives? Two car garage or three?
- Back yard / covered patio's / porches / Landscaping / Pool /

These are just a few things to consider when you start to think about the design and layout of your new home. I would strongly recommend you visit several open houses in your area and make note of the features of each home that you like, add theses to your list. Also, review as many home planning books that are available and make note of the features you and your family prefer in a home. Take all of this information and sit down with a residential design professional, either a residential architect or an architectural design draftsman and discuss your ideas. They can put together a rough draft of what they see your new dream home to be.

Working with a design professional will help take some of the stress off you as you go through the process and should help you get some solid input on the design of your new home. Be sure to have the designer produce a complete set of construction documents including;

- Site plan, showing building location on the lot with all drives and walks
- Floor plans showing room layouts and sizes
- Exterior elevations showing roof lines and exterior finishes
- Mechanical / electrical & plumbing plans showing lighting & switch layouts
- Specifications sheet showing specific models and colors of equipment & finishes

Then have enough copies of plans to give to your subcontractors pricing the project.

Estimated Costs / Securing the Funds for the Project

After your plans are completed, you may wish to take your complete set of construction documents including specifications to your lender to have them conduct an appraisal of the home's value based on the blueprints. This is one of the steps they will need to take in order to secure your construction loan to begin the project.

You may also wish to conduct your own survey of the costs prior to seeing your lender. On the following page is a sample of a spreadsheet called Schedule of Values for Construction. This sheet lists the various tradesmen involved in constructing your project and the amounts of their subcontracts. It is always a good idea to obtain at least three prices for each trade, in other words, put the project out to bid to three subcontractors for concrete work, framing, roofing, electrical, plumbing and all the other work of the project. By doing this you will not only be sure you are getting fair pricing, but when you are done you should have a feel for the complete cost of the project also.

Shop several lenders for the best interest rate and terms for your construction loan. Let them know they are competing with other lenders so they will get serious about giving you the best terms. Keep in mind that if you do get a construction loan, you will begin making monthly payments on the amounts used each month in most cases. In other words if you spend $30k the first month, that is what your payment will be based on the second month, spend a total of $80 by the second month and your construction loan payment increases throughout the course of your construction project. When completed, you will have used all of your construction funding and your monthly payments will be based on your overall project costs, converted to a standard mortagage.

You may also find it to you benefit when setting up the funding for the project to have all of your funds dispersed through a local title and escrow company. This will insure that when funds for completed work are released to the subcontractor, all responsible parties in control will have an opportunity to review the subcontractors request for payment and verify that he is being paid only for the amount of work completed. You, your lender (banker) and the escrow company will have opportunity to review and approve each request for payment. The cost of the escrow is minimal and well worth it.

Schedule of Construction Values Spreadsheet

You may wish to track your job costs using a simple excell spreadsheet known as a Schedule of Values for Construction. By doing this, you will be able to identify the key trades which controll the cost of the overall project and will be able to see where your money is going. See the sample chart below;

| SCHEDULE OF VALUES - BY CATEGORY | | | | DATE: | |
DESCRIPTION OF WORK	SUBCONTRACTOR	MATERIAL	LABOR	EQUIPMENT	TOTAL
					$ -
Clearing & Grading	Joe's Excavation	$ -	$ 2,500.00	$ -	$ 2,500.00
Excavation & Dirtwork	Joe's Excavation	$ -	$ 4,000.00	$ -	$ 4,000.00
Temporary Utilities / Toilets	Jonny on the spot	$ -	$ -	$ 450.00	$ 450.00
Foundations	Bobs Concrete	$ 3,500.00	$ 2,650.00	$ 1,200.00	$ 7,350.00
Concrete Walks/ Drives/ Porches	"	$ 2,000.00	$ 950.00	$ 120.00	$ 3,070.00
Concrete Slab on Grade	"	$ 4,000.00	$ 1,500.00	$ 800.00	$ 6,300.00
Rough Carpentry Framing	Tri State Framing	$ 24,000.00	$ 8,500.00	$ -	$ 32,500.00
Exterior Finish Siding	Finishes Inc.	$ 8,500.00	$ 3,950.00	$ -	$ 12,450.00
Roofing	Jacks Roofing Co	$ 4,500.00	$ 4,500.00	$ 450.00	$ 9,450.00
Exterior Doors & Windows	Lowes	$ 12,000.00	$ -	$ -	$ 12,000.00
Thermal Insulation / walls - ceilings	The insulation guys	$ 3,250.00	$ 2,450.00	$ -	$ 5,700.00
Sheetrock / Walls & Ceilings	Hermans Drywall	$ 5,500.00	$ 3,250.00	$ 1,200.00	$ 9,950.00
Interior Doors & Trim	Lowes	$ 4,200.00	$ 3,200.00		$ 7,400.00
Interior Painting	Blues Painting	$ 2,650.00	$ 4,000.00	$ -	$ 6,650.00
Exterior Painting	Blues Painting	$ 3,650.00	$ 4,000.00	$ -	$ 7,650.00
Kitchen Cabinetry & Countertops	Lowes	$ 12,500.00	$ 4,500.00	$ -	$ 17,000.00
Electrical Wiring & Fixtures	Empire Electrical	$ 8,500.00	$ 3,600.00	$ -	$ 12,100.00
Plumbing Piping & Fixtures	AAA Plumbing Co	$ 4,500.00	$ 2,300.00	$ -	$ 6,800.00
Heating & Air Units & Ductwork	Mikes Heating & Air	$ 9,000.00	$ 3,250.00	$ -	$ 12,250.00
Misc Cabinet Hardware	Home Depot	$ 450.00	$ 300.00	$ -	$ 750.00
Finish Floor Coverings	Home Depot	$ 6,500.00	$ 2,500.00	$ -	$ 9,000.00
Closet Shelving & Special Millwork	TBD	$ 1,250.00	$ 800.00	$ -	$ 2,050.00
Appliances	TBD	$ 11,500.00	$ 1,650.00	$ -	$ 13,150.00
Special Finishes	TBD	$ 2,300.00	$ 1,200.00	$ -	$ 3,500.00
Landscaping	TBD	$ 4,595.00	$ 2,250.00	$ -	$ 6,845.00
Continuous Cleanup / Dumpsters	TBD		$ 650.00	$ -	$ 650.00
Permenant Utility Hookups	TBD		$ 250.00	$ -	$ 250.00
					$ -
Contingency Budget	Self	$ 10,000.00			$ 10,000.00
Total Projected Cost					$ 221,765.00

Many variations of this spreadsheet can be used to help you track the costs of your project. It is helpful to list the description of work, the responsible party and total costs. Throughout the course of the project you will be asked to release payment for work completed. Establish the percentage of work complete, and pay only this percentage of the total amount to your subcontractor. You want to hold sufficient funds at all times to insure completion of each trades work.

Documentation / Permits / Insurances

Documentation

Minimum documentation required to start the construction project could include not only your complete blueprints & specifications but also a survey / plot plan which shows a legal description with easements, setbacks and property lines. Your survey drawing should show the locations of corner pins and existing site utilities which will be helpful when deciding where to locate your home on the property.

You will need to check with the local building authority which issues the permit, as well as any homeowners association regarding minimum setback requirements. There may be minimum setbacks for the front and sides of your property, in other words you cannot build within so many feet of the property line. The front setback gives uniformity to the entire development, establishing the building line along the street.

Also this site plan / survey in locating any existing utilities such as electrical, water & gas lines will aid you in locating where your homes utilities will tie in. No fencing or other structures should be constructed on or over the utility easements.

Your construction documents should include a full set of blueprints with the following sheets.

Site Plan – locates the structure to be built on the property along with any drives, walkways, accessory structures & location of utility which you will hook up to.

Foundation Plans & details – shows the footings, foundations and reinforcing layout

Floor Plan – Shows room layout and configuration with dimensions and room sizes

Exterior elevations – Shows all sides of the exterior of the home with finishes

Framing Plan – a layout of the structural components for floors, ceilings walls & roof

Electrical Plan – room layout showing the locations of all lights, receptacles & switches

Specifications sheet- detailed information on interior finishes, paint colors and textures, trim and door types, special coatings, electrical & plumbing fixtures, window and door types, hardware, and specific model numbers of equipment. And anything else you can think of that is specific to the project.

Permits

If you are building in an area that is regulated by a governing authority, either city or county, you may be required to apply for one or more permits prior to beginning your construction project. Various types of permits are often required, and each comes with a fee and usually inspections for approval from the issuing authority. Some of the more common permits you may be required to apply for prior to beginning are;

Clearing & Grading permit (land disturbance) to clear the property

Foundation Permit - for installation of underground footings, foundations & slabs

Building Permit – for the general work of the contract

Electrical Permit – for installation of all electrical components & systems

Mechanical Permit – for installation of all heating & air conditioning systems

Plumbing Permit – for installation of all plumbing, sewer & water systems

Storm Drainage Permit – for installation of site drainage systems

Landscaping Permit – for installation trees, shrubs , fencing & green areas

Occupancy Permit – Documentation of final inspection and right to occupy

Some locations have one general permit for construction while others have more specific breakdowns as shown above – requiring many of your subcontractors to pull the individual permits for their particular trade. Be sure to check with your local building authority prior to beginning your project. They can give you a schedule of all required inspections and the lead time required for calling in the inspection request. It is in your best interest to cooperate fully with the Building Regulations department and inspectors as they are working in your behalf to insure a safe and well constructed project. Also please note that normally you will be required to submit a complete set of construction documents for their review when applying for the required permits.

Insurance

Contractors Liability & Workman's Comp Insurance

Be sure your contractors or subcontractors have both general liability and workman's comp insurance. Make it a requirement of the subcontract agreement that they have these policies currently in place prior to starting work on your project. Generally speaking, you will get a better class of subcontractor if you hire one who already carries the minimum insurances. Workman's comp insurance will compensate any worker who is injured on the job by covering lost wages and medical costs due to an injury sustained while working on your project. It protects both you and the employee.

Builders Risk Insurance

Builders Risk Insurance is a special type of property insurance which typically covers loss of materials, fixtures & equipment during construction. Common Builders Risk coverages may include loss from fire, theft, vandalism & storm damage. Most Builders Risk policies do not cover losses from earthquake or floods, and some policies near coast lines will not cover loss from wind damage. Check with your local provider for a specific breakdown of what your policy does and does not cover. Know the facts. It is common for the property owner to carry this coverage, however, on occasion the Contractor will provide this coverage as a part of their overall service. Near the end of your construction project you will need to convert to a complete homeowner's policy for the long term. You will need to coordinate the start of this permanent coverage with the termination of the Builders Risk policy to insure you have no loss of coverage at any time.

In some cases of remodeling, the renovations may be covered under your current homeowner's policy and a Builders Risk policy will not be required. Consult your insurance agent for professional guidance in this matter.

Subcontractors & Subcontract Agreements

When interviewing subcontractors to work on your project, you will want to give each one a complete set of construction blueprints and specifications. This way you will insure that all the pricing of the work is based on the same information. Always be sure to get pricing from at least 3 sub's from each trade, which may include the following;

Earthwork / Grading Sub – clears brush & trees, brings building site to subgrade

Concrete sub – for footings, foundations, basements, slabs, drives & walkways

Framing sub – General framing of the home, sets doors & windows, dry in the roof

Roofing sub – Installs roof felt, flashing, shingles, skylights , drip edge,

Gutter sub – Installs guttering & downspouts / sometimes exterior siding

Electrical sub – Underground electric, rough in electric, hook up to all appliances, finish trim out of electric and installation of all electrical & lighting fixtures. Also provides temporary electrical service for use by all other subcontractors during construction.

Plumbing sub – Hookup to all existing water & sewer services, rough in plumbing, all plumbing trim out , setting of plumbing fixtures and hookup to appliances such as dishwasher & refrigerator icemaker.

Mechanical sub – rough in all heating & air conditioning ductwork, thermostat wiring, line sets, condensate lines, heating & air trim out of grilles & registers.

Insulation sub – Installs all thermal insulation for attic spaces, floors & walls

Sheetrock /drywall sub – installs all drywall , tapes, muds and sands for prep of paint

Painting sub – primer coat, painting of all walls, ceilings, doors & trim

Flooring sub- furnish & installs all floor coverings & preps floors to receive all flooring

Millwork / trim sub, - installs all base trim, wall & ceiling trim, sets cabinets & hardware

Misc installations – towel bars, closet shelving & rods, misc hangers

Landscaping sub – installs all exterior plantings, sod, grass & fine grading of site

Subcontract Agreements

You've heard it said, "Agreements prevent disagreements." Nothing could be truer than this when it comes to construction. Having your work agreements in writing is a must when you begin your project if you intend to be serious about controlling your costs. Most subcontractors are hardworking honest folks, but there are a few out there who play on the inexperience of the do it yourselfer and here's where my experience comes in. Your subcontract agreements need to be as complete as possible so you don't get blindsided by unexpected costs. Here are a few things you can include in your various subcontracts to insure a more complete project;

Clearing & Grading; be sure the subcontractor includes a detailed list of what they will do for the price. Clearing a number of trees, removal of the roots, haul off of demolition debris, cost of permits etc.

Concrete ; subcontractor should include the haul off of spoils – in other words if your concrete subcontractor digs a basement or foundations they need to include in their pricing to haul off the unsuitable / unusable soils from the dig, otherwise you will end up with a large pile of dirt you have to haul off at your own expense. Include all materials for a complete project including gravel fill, vapor barriers, reinforcing rods, anchor bolts, curing compounds, sealants, removal of forms, and removal of unused concrete materials (from cleanout of the concrete truck), all taxes on materials and insurance on workers included in the contract costs.

Framing subcontractor – include all materials for a complete project, dimensional lumber, plywood, nails, screws, bolts, miscellaneous fasteners, lifting & unloading equipment, and all equipment & labor for a complete installation. Note; if the framer does not include the framing lumber in his contract, you could see excessive waste of materials and additional costs, so it is a good idea to have them include all materials in their price. You may ask for a breakdown of labor and materials, and then have other lumber suppliers give you a bid for materials only. You can then use this material price to insure your framing subcontractor has indeed included all materials for the project.

Roofing subcontractor; needs to include all labor materials and equipment to install roofing felt, shingles, ice shield, drip edge, valley flashing & roof cap. This includes all nails & fasteners, lifting of materials to the roof and especially cleanup of all debris.

Clearing / Grading / Earthwork and Layout

Before starting your project there may be some work that needs to be done to the site to make it acceptable as a building location. Trees may need to be removed as well as old structures or underground utilities in some cases. You will need to walk the property with your excavation (dirt) subcontractor and discuss where you wish the home to be built, locate any drives and out buildings.

Also look at the grade / slope of the property for drainage. Some reconfiguring of the elevations may need to take place for proper drainage. Then picture what you would like your yard to look like when the project is finished, with any unsightly trees and underbrush removed.

- Walk the project with your subcontractor and discuss the building location
- Mark with surveyors tape, any trees and shrubs you wish to keep
- Look at what trees or shrubbery needs to be removed to improve the view

Your subcontractor can now take your building plans and begin the layout of the home, and drives. He will need to bring the building site up to what is known as "subgrade" in other words bring the building pad up to a level that your next subcontractor who works on the footings and foundations can take it from here without a lot of additional grading and dirt work.

In some instances your site will need soils hauled in to bring the building pad up to a higher elevation prior to the construction work starting. In this case your subcontractor will bring in the fill and compact it in what is known as "lifts." The fill materials will be brought in in layers, each one compacted to about 95% so that your building pad is stable. You will want your new built up building pad to be somewhat larger than the actual footprint of your home, this will allow for the foundation work to take place without difficulty. If you don't need fill, but instead have to "cut" or remove elevations of the soil to level the building pad, ask your subcontractor to include in his bid, a cost of removing spoils or hauling off the unsuitable soil.

Site Utilities / Temporary Utilities

Site utilities are the electrical , water, sewer and drainage systems outside the footprint of your home, the utility lines leading to and connecting main line at the street. It is important that you do some research prior to purchasing your property (see Chapter 2) to determine which of these utilities are already in place either on the property or if they are part of a utility easement nearby. If they are not already in place, you may have to bear some of the cost for the extension of these utilities to your property.

Electrical service lines are still run overhead in many locations and are thus obvious. Some developments have all underground utilities and you will note that transformers are located throughout the development. You will need to check with your local utility supplier and see if your property can be serviced from one of the existing transformers.

Water service lines may run near your property (water main) but no actual line running to your construction site. In this case you will need to discuss with your plumbing subcontractor the cost for extending the line, installation of a water meter and shut off valves and hookup to your home . This is known as "tie in" to the utilities. Your water service company may also require deposits to hook up to or turn the meter on, so check with them to get an idea of all the costs.

Sewer systems – City sewer mains are either gravity type or force mains (pressurized). Check with the local utility company and see which type runs nearest your property and have your plumbing subcontractor include the cost of hook up or tie in to this utility in his overall scope of work. If there is no existing sewer service near your property, you will need to install an independent septic system. Regulations vary across the country for the installation of systems so check with your local city or county offices to be sure of the requirements. Basic requirements would include a "perk test" of the soils to see the capability of the soil to handle the amount of sewage based on the size of the home. This will also govern the amount of "field lines" installed to distribute the waste over a determined field or area. You will need a quote from a certified septic installer for a complete turn key project if this is the case, along with complete drawings to submit for review to the approving government offices who regulate the installation.

Temporary Utilities – Discuss with your electrical subcontractor and your plumbing sub and have them include in their contract, the cost of temporary electric and water for use during the construction project. Simple enough, but very necessary for a smooth project.

Concrete Foundations / Walks & Drives

Now that your building site has been cleared and the corners of the new home located, you are ready for the footings and foundations. Your concrete subcontractor will need to set building lines, establishing all the inside and outside corners of the home's exterior perimeter. If your home does not have a basement, then your subcontractor will proceed with the digging of perimeter footings and any thickened footings within the footprint for load bearing support. Your architectural designer who provided the construction plans should also provide details (cross sections) of the footings indicating their width, depth and size & spacing of reinforcing bars required in each footing. It is important to note that when the reinforcing bars are tied in place and ready for concrete, none of the reinforcing bars should be in contact with the earth but instead should be supported by "chairs" or bricks. Proper support of the reinforcing bars is important to hold them in their assigned place during the concrete pouring process.

Across the country building code requirements for depth and size of foundations & footings will vary depending on the frost depth for that area. In all cases, bottom of footings need to be located on solid undisturbed soils beneath the frost depth shown in the building code for that region. Placing the footing below the frost line prevents movement of the footing during the freeze/thaw process.

Sleeves through the foundation will need to be placed prior to pouring concrete to allow for the passage of any utilities such as electrical, water & sewer service. These sleeves will need to be coordinated with the site utility plan and the floor plan of your home. For example, look on your homes floor plan and see where the water service is shown to enter the home (likely in a utility room). This water service will be run under ground or under your concrete floor and the water line will exit the home through a sleeve in the footing. Locate the sleeve in a convenient location to hook to the water service at the front of the property. The same principle will apply for electrical service and sewer service.

Your drives and walks will be a bit less critical since they are not a part of the occupied structure. Most drives are based on 12' width per vehicle, so a drive for 2 cars would be 20' to 24' in width, walkway width is more of a personal preference says 3' to 5' in width. In laying out your drive approach, also consider any turn around areas you may need.

Basement Foundations & Walls

If your home plan calls for basement construction, you will also need a full set of details and cross sections on the plans to show clearly all the various things required. Much more extensive earthwork is required when installing a basement foundation. Most of the earth /soils removed for the basement will not be suitable for anything else during your construction project so ask your subcontractor to haul the excess soils off your site.

Basement construction, when done right will include not only waterproofing of the walls and foundations which will be subject to backfill, but also installation of foundation drains around the perimeter. A premium product that I like to recommend for waterproofing the basement walls subject to backfill is called Paraseal by Tremco. The product is a rubberized HDPE sheet with a coating of Bentonite clay on the backside. This provides two layers of protection, first the rubberized sheet repels moisture and the bentonite clay swells up to 8 times its size if moisture passes thru the membrane. The properties of the clay swelling, seals off any water trying to penetrate into the foundation. Proper installation also requires powdered bentonite to be installed at the cold joint where basement footings and walls come together, making a complete water proof seal. When this system is properly installed and backed up with a drainage system, you will enjoy a basement without any water or moisture issues.

Drain piping should be installed around the perimeter of the footing and backfilled with clean rock for a distance of at least 2 feet from the basement wall. This will allow any ground water to drop through the clean rock and into the drain, carrying it away from the building and not creating a standing water issue against the wall.

Consult with your architectural designer for specific construction details on your basement. The more details your subcontractors have at their fingertips, the fewer questions and issues will arise during construction. The subcontractor will need specific information also on size and placement of all reinforcing bars in the basement walls and foundations.

Rough Framing Carpentry

Rough framing carpentry is just that, it is the rough frame for your home's exterior walls & roof and interior walls, floors & ceilings. It includes other subframing such as fur downs or chase ways for heating , staircases and blocking or backing for other items.

If you decide to construct your project on what is known as a crawl space, you will have foundation walls which support cross beams or girders which in turn will support all of the floor joists and floor decking.

Note in this photo, the main floor beam or girder running through the center of the home and is supported by individual piers or posts at designated intervals. This girder or support beam is an engineered wood product capable of supporting the entire floor framing system. The workers in the photo are beginning to install pre-engineered floor joists which run the entire width of the home. Note that the joist are bearing on pressure treated lumber on the exterior concrete foundation walls.

Installation of all pre-engineered floor joists prior to the subfloor plywood decking

Rough framing includes the framing of exterior wall & roof and wall & roof sheathing

Roofing Materials & Drying In

When rough framing of your home nears completion, the framing crew will install roof decking made either of plywood sheets or individual 1x lumber spanning the rafters. The framing subcontractor will normally include in their contract to install the roof decking and sub fascia, which is sub framing for the fascia's vertical face.

You will need to schedule your roofing subcontractor to start installation of the roofing materials immediately after the framing subcontractor has installed the decking. You don't want the decking exposed to any wet weather while it is unprotected. The roofing subcontractor will first install roofing felt or deck armor to protect the wood decking and provide a moisture barrier between the shingles and the wood. Whether you choose to use regular roofing felt, deck armor or another product is your choice. Here are a few things to remember when hiring your roofing subcontractor;

Be sure your roofing subcontractor has included labor, materials and equipment to install all of the items listed below in his price for a complete project. The first time I had hired a roofer, he simply gave me a price per square to install the shingles on the roof. Later when I got the bill it was several hundred dollars higher because he claimed that all the miscellaneous items were not included in his original price. So be sure to include the following in your subcontract with the roofer;

- All labor materials and equipment for a complete roofing installation
- All felt, underlayment, deck armor, and starter shingles
- All valley flashing, step flashing, flashing of openings, roof boots
- All drip edge and starter shingles, ridge cap shingles
- All lifting of materials, men and equipment to the roof
- All nails, screws, staples, fasteners, adhesives, sealants and caulking
- All crickets if required for positive drainage
- Ridge vent, soffit vents or prep for soffit vents,
- Cleanup and haul off of all construction debris, scraps and materials
- Liability and Workman's Compensation Insurance for the project

If you carefully include these items at a minimum, your roofing project will go much smoother. Refer to the following page for a comparison of various types of roofing materials.

Roofing Material Types – Comparison

Just as in any other part of your home construction, when it comes to roofing materials you have a wide variety of choices. From the felt underlayment to the finished shingles on top, there are several products to choose from. Here is a bit of information to think about in comparing materials.

Roofing felt is a basically a rolled felt material saturated with asphalt which is used as a protective barrier under the shingles and on top of the wood roof decking to protect it from moisture that may migrate through or under the shingles, say in a storm. The issue with roofing felt is that its moisture resistant quality is somewhat weak, and when exposed to the elements it will actually hold moisture and swell, causing extensive wrinkling and does not help give you a smooth roof. The felt comes in different weights starting with 15 pound and 30 pound as the most commonly used. The weight reflects the amount of asphalt saturation and paper thickness. Standard roofing felt has been used successfully for years and is a good product if covered immediately by the final roofing shingles. However, improvements in products continue to be made and one that is exceptional is called Deck Armor. Similar to roofing felt, but about one third of the weight, and it can be left exposed for days and will still lay flat and true. While it is water resistant, it is not waterproof, but it does allow the home to breath.

Valley flashing is a product put in the area of your roof where two roofs come together and form a "valley". It protects the wood decking from water entering at this vulnerable point of the structure. Starter shingles are a course of standard shingles laid upside down along the lower edge of your roof, all around the perimeter and give your regular shingles something waterproof to adhere to. Step flashing will be found along the edges of say a chimney or other vertical edge coming thru the roof, and it is in short pieces laid successively on top of each other forming "steps" so the water from above always runs off on to the step flashing below. Any pipes venting the plumbing systems will need "roof boots" which consist of rubber boots to snuggly fit around the pipe and have a base of flashing built in to be installed during the shingle installation.

Roofing shingles are normally called "three tab" (flat shingles) or "architectural " shingles which have a raised textured finish. Most three tab shingles will come with a 25 year to 30 year limited warranty and architectural type will range from 30 to 50 year. Check with your local building supply house to see the various types before choosing.

Thermal Insulation of Walls / Ceilings / Floors

Thermal insulation of your home needs to be done properly if it is to perform as designed. There are many products on the market now and all of them seem to have their good and bad features so again, do your homework before deciding on which type of insulation products to use.

Fiberglass batt insulation has been a popular choice for many years and if properly installed, it will provide you with a comfortable home on the coldest winter nights or cool in the hot summer. Fiberglass batt insulation comes in various widths based on the spacing of the framing members found in your home. If your floor joists are spaced on 24" centers you will need insulation batts 23" wide. This will allow the insulation to fit in between the framing members tightly for a good fit. If your wall studs are on 16" centers you will need 15"wide batts.

In wall / floor /ceiling insulation the thickness of your framing will govern which type of insulation you will use. If you have 2" x 4" framing in the walls you will use an R13 batt and if you have 2" x 6" framing you will need an R19 batt which is thicker. Floor and ceiling framing may be 2" x 12" allowing an R30. The insulation batts only perform well if installed properly. They need to be a snug fit but not smashed into the space as you may have seen. The batt insulation is designed to trap air between the fibers and this can't happen if it is forced into the wall too tightly. The same goes for ceilings and floors. Batt insulation in the ceilings and floors is always a challenge to keep it in place. The weight of the insulation batt pulls it from its place if not secured properly. Wire hangers are sometimes used in flooring applications as is netting or small rope. Wall installations can be stapled into place using the paper tabs on the sides of the insulation batts. In any case it needs to be carefully installed around any receptacle boxes or wall switches, taking care that all voids are filled to prevent air infiltration.

Spray foam insulation is gaining popularity due to its terrific insulation ability. You will achieve a higher R-value in the same wall or ceiling cavity with spray insulation. While this type of insulation is gaining popularity it too has its issues. Some cases the spray foam will shrink and pull away from ceiling joists or wall studs creating a cold spot.

So before choosing a product, speak to your insulation subcontractor and discuss the options.

Punch Lists / Final Payments / Lien Releases

Congratulations, several areas of your project are now complete. Now is the fun part, finishing up! It is common on all construction projects to walk the entire project and make a corrections list called a "punch list." Your subcontractors will know what you are talking about when you mention this term, and you don't have to wait until the end of the project to make the list for most sub's, just the ones associated with the finishes.

Walk through the project when you see it is about finished up and make a list by room or area that addresses anything that just doesn't feel or look 100% correct. For example;

- Door sticks when shutting or doesn't close completely
- Cabinet door needs alignment
- Cracked floor tile in laundry room
- Paint runs on door frame or wall paint spilled on trim

Anything and everything that you want fixed needs to be put on the list, and then a copy given to the party responsible for the work and the repair. You will want to do this early on for your framing and concrete subcontractor as their work will be covered up by other trades.

Another good idea when making payments throughout the course of the project is to have each subcontractor sign a waiver of lien when they receive their check for the work. This will keep you from having to pay for the work twice as it is a legal document stating they have paid for the labor and supplies used during this billing cycle. Earlier in this document I recommended you hire an Escrow agent to handle disbursement of all payments of funds for the work. If you do, they will require the subcontractors to sign the waivers when they pick up their checks. Hiring the escrow staff to handle this takes a lot of the stress off you the homeowner and keeps everything running as budgeted. The laws regarding lien release, lien waivers, and material liens vary depending on your state, so check with your local title and Escrow Company for the specific requirements in your area before starting.

I hope this publication has enlightened you to the fact that there is much to think about when starting to construct a home.

New home framed and ready for brick exterior

Rough Frame Exterior –not dried in

Other publications by this Author

Available on Amazon –Kindle ebooks

- Basic Project Management for Commercial Construction
- Lose 40 Pounds in 8 Weeks – The Easy Way
- Finding My Way Home – An Alzheimers Story
- The World According to Roy – a humerous look at life

www.ingramcontent.com/pod-product-compliance
Lightning Source LLC
Chambersburg PA
CBHW041310180526
45172CB00003B/1050